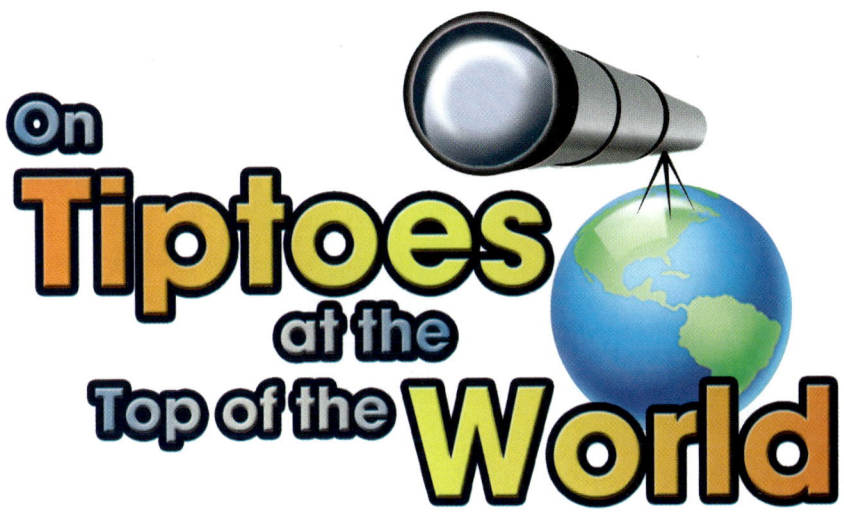

On Tiptoes at the Top of the World

by Mark Fara

SCHOOL PUBLISHERS

Orlando   Austin   New York   San Diego   Toronto   London

Visit *The Learning Site!*
www.harcourtschool.com

Introduction

Suppose that it is thousands of years ago, long before skyscrapers or streetlights. Your home is a cave, and the only light in your cave comes from a fire. You never flip a light switch or see a light bulb because electricity has not been discovered yet.

It is nighttime, and you are bored. You decide to take a walk outside. You take a torch with you to light your way. But as you reach the edge of a nearby field, a breeze blows out your torch.

At first you are nervous about suddenly being in the dark. But then you look up. The sky looks like an enormous black dome filled with bits of light. There are no clouds, and no moon is visible. But there are so many bits of light in the sky that you could never count them all. Your eyes adjust. The longer you look, the more you notice. They are stars, of course. But there are other objects, too—faint, hazy somethings that look like faraway clouds. Now and then you see something bright shoot across the heavens and then vanish. You might see something that looks like huge illuminated bedsheets flapping slowly on an invisible clothesline. This is what people now call the aurora polaris, or polar lights.

Outer space is fascinating. For thousands of years, people have wondered what is out there and have tried to find out. Some people have built buildings called observatories to help them watch and learn about the stars. The scientists who use them are called astronomers, and the science of studying things in the sky is called astronomy.

Today, astronomers have many different technologies for studying objects in space. They use X-rays, radio waves, infrared rays, and ultraviolet rays. Scientists send satellites into orbit to collect information and send probes to distant planets to do experiments and send back data. Even with all of these new technologies, however, astronomers often rely on the original method for observing the sky: looking at objects with visible light. Most satellites and space probes have cameras that take photographs in visible light and send them back for us to see. Most observatories have telescopes that allow astronomers to look at the sky with their own eyes or set up cameras to take photographs for them.

Astronomers who use visible light to study space are called optical astronomers, and their job is called optical astronomy. They do everything they can to get the best possible view of things in space using light. You might say that they stand on tiptoe at the top of the world, trying to see farther into space than anyone else ever has.

Aurora polaris, or polar lights, are caused by particles from the sun entering our atmosphere. If they are seen in the north, they are called aurora borealis, or northern lights. In the south, they are called aurora australis, or southern lights.

The First Optical Astronomers

Until very recently, all astronomers were optical astronomers. You may have heard of a place in England called Stonehenge. Did you know that many archaeologists think that Stonehenge was an ancient space observatory? The stones of Stonehenge were arranged in a pattern that helped people predict eclipses and other events related to Earth, the sun, and the moon. We may never know who built Stonehenge, but whoever it was knew something about astronomy more than 5000 years ago.

Those who built Stonehenge were not the only ones to notice that the sun, the moon, and stars seem to move across our sky in ways we can predict. Babylonians began charting the stars at least 3800 years ago. The people of ancient China also studied and learned about objects in space.

Ancient philosophers thought about stars and wrote down their ideas. Rulers of kingdoms built observatories. There was even a school of astronomy in Baghdad in the early 800s.

About 2500 years ago in Greece, a man named Eudoxus of Cnidus built models of Earth, the sun, the moon, and some of our neighboring planets. He discovered that our year is 365 days long, with 6 hours left over. Thanks

Historians believe people used Stonehenge to predict eclipses.

to his discovery, we have a year with an extra day (a leap year) every four years. This extra day makes up for four years' worth of extra hours. Otherwise, every year each day would occur a little earlier. We would eventually have summer in December and winter in July.

Constellations

Have you ever looked at the clouds and thought that some of them had the shapes of faces, animals, or objects? Ancient people did the same thing with stars. They divided the night sky into groups of stars that formed different shapes. Groups of stars that seem to form a shape are called constellations. Most constellations have myths that people tell about them. A myth is a story to explain something about the nature of the world or of people. Myths are told and retold until they become a part of a culture's way of experiencing life.

As far back as 6,000 years ago, people in parts of the Middle East had identified constellations. We know this because they drew pictures of the constellations on everyday objects. The Egyptians, Greeks, and Romans came up with their own constellations. In 1930, a group called the International Astronomical Union decided on the names and boundaries of 88 different constellations.

The Big Dipper

5

The most famous constellations and the easiest ones to spot are probably the Big Dipper and the Little Dipper. At least two different cultures associated these constellations with bears. The Greeks believed that the Big and Little Dippers were a mother bear and her son who had been tossed into the sky by Jupiter, the king of the gods. In America, a Native American tribe called the Navajo believes that the seven stars making up the Big Dipper are seven brothers who flew up to the sky to escape their angry sister, who could turn herself into a bear.

Another famous constellation is Orion the Hunter. The ancient Greeks had many myths about him. Some said that he was the son of a poor shepherd. Others said that his father was the sea god Poseidon and that his mother was a Gorgon, an ugly flying woman with snakes for hair who turned people to stone when she looked at them. Some said that Orion angered a goddess by hunting her animals and was killed by her giant scorpion, Scorpio, which is also a constellation. Others said that Orion was blinded by his girlfriend's father and was later killed by a jealous goddess. All of the stories agree that Orion was a great hunter and that he was placed in the sky as a constellation.

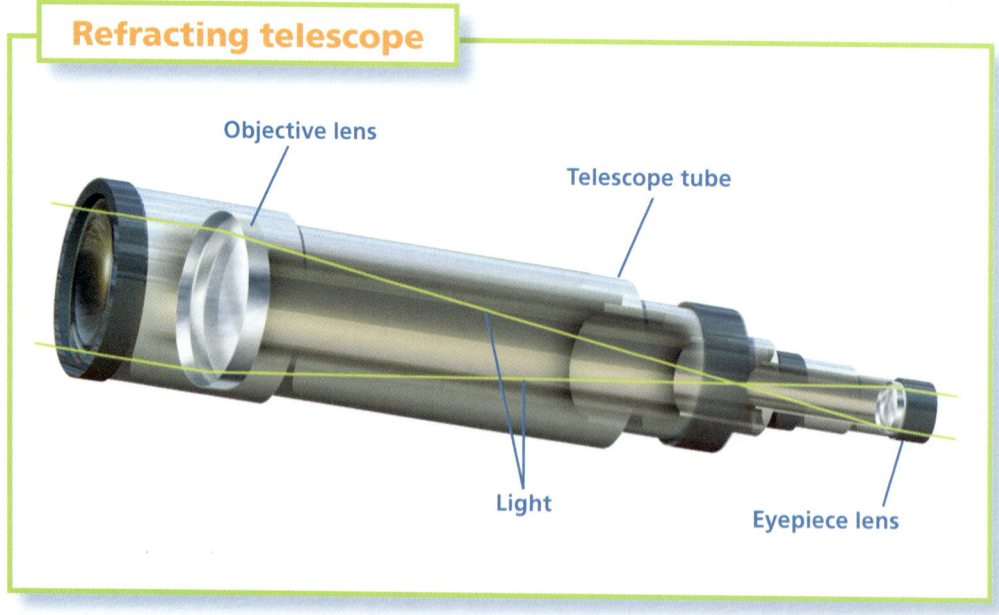

Refracting telescope

Objective lens

Telescope tube

Light

Eyepiece lens

Telescopes

As scientists learned more, they realized that they needed better ways of looking at objects in space. In the early 1600s, Europeans made the first telescopes. A telescope is an instrument that makes objects in the sky look nearer and larger. You can buy small telescopes in stores. Much larger telescopes are used in today's space observatories.

The earliest telescopes were called refractors, or refracting telescopes. An Italian named Galileo Galilei was most likely the first person to use a telescope to look at the sky. He built his refractor in 1609. He used two lenses that made the objects being observed look larger. One of the lenses was large and the other was small. The lenses were mounted at either end of a long tube. Galileo put the small lens up against his eye and pointed the large lens at the object he wanted to observe.

Refractors are still used today, but two major problems limit their use. One problem is that glass is very heavy. It is hard to make lenses large enough to magnify enough to be useful to astronomers. Another problem is that what the human eye sees as white light really contains different colors, and lenses often make these colors visible. As a result, a lens makes objects appear to have different-colored rings around them.

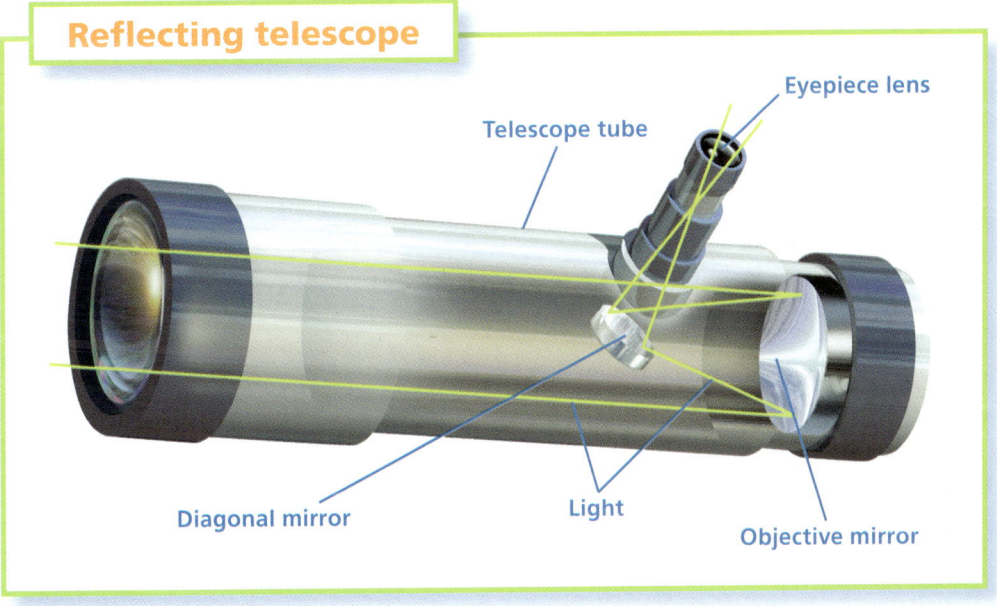

Reflecting telescope

Eyepiece lens
Telescope tube
Diagonal mirror
Light
Objective mirror

In 1668, an Englishman named Isaac Newton built the first reflecting telescope, or reflector. A reflecting telescope uses a magnifying mirror instead of the large lens of the refractor. The person using this type of telescope looks at a reflection of what is being observed. A good mirror does not separate white light into colors. It is also easier to make large mirrors than it is to make big, thick, heavy lenses. Most large telescopes built today are reflectors.

Observatories

Observatories are places built especially for observing the night sky. A good observatory cannot be built just anywhere. Observatories are usually not built close to cities, for example, because of the city lights and hazy city air. A dry climate is also important because moisture in the air interferes with observations. California and New Mexico have observatories. These locations are good because they have little rain and few clouds.

Most observatories are on high mountains. That is because the higher the observatory is, the less atmosphere you have to look through to see objects in space. Even though air is invisible, it can affect light from outer space. If you look at the stars at night, you can usually see them "twinkle." The stars are not really twinkling, but air makes them look as if they are.

At many observatories, a dome protects the telescopes from wind and bad weather. Astronomers can open a slit in the dome to view the sky at night. The dome can rotate so that astronomers can view any part of the sky.

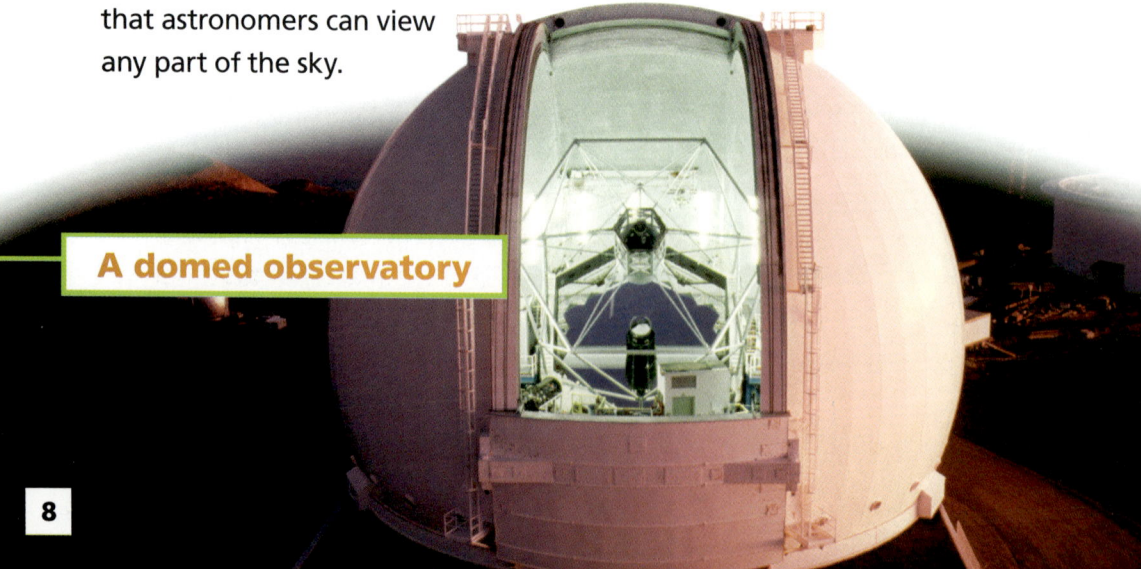

A domed observatory

Some observatories are small buildings from which the entire roof can be removed. This kind of observatory does not protect telescopes as well as a domed observatory does, but it costs much less to build. Therefore, amateur astronomers, people who enjoy astronomy but do not make their living at it, usually build this kind of observatory.

Sometimes optical telescopes are used as cameras. Many objects in the sky are so faint that they cannot be seen by looking through the telescope. However, if the telescope is focused on the object for several hours, enough light can reach a piece of film in the telescope to produce an image. The telescope is set up with a motor that keeps it pointed at the object as the object moves across the sky. Otherwise, the object would show up on the image as a faint line instead of as a clear picture.

Some observatories are set up to take photographs automatically on every clear night. Cameras in this kind of observatory provide us with pictures of meteors, asteroids, comets, and other objects that move in our skies. Other observatories take pictures during the day. These observatories take photographs of the sun to keep track of the movements on its surface. The information from these observations is important because activity on the sun can affect radio and television signals here on Earth.

Because it is impossible to see the whole sky from any one place on Earth, it is important for observatories all over the world to share their observations and combine their information.

Some Interesting Observatories

- The most powerful telescope in the world is located at an observatory high on a mountain in the Atacama Desert in Chile. This is a particularly good place for an observatory because the Atacama Desert is one of the driest places on Earth. The telescope is called the VLT, which stands for Very Large Telescope. The VLT is really four large telescopes plus a few smaller ones. Images from all of these telescopes are put together to make larger images. Plans for this facility include an even larger telescope, which will be called the OWL (OverWhelmingly Large).

- In 1948, the Hale telescope began operation at the Palomar Observatory in California. It was named after George Hale, an astronomer who studied the sun. The Hale telescope was the largest single telescope in the world until 1974. A bigger one was built in Russia that year, but a flaw in its mirror keeps this telescope from being very useful.
- In 1993, two telescopes were installed on Mauna Kea, an inactive volcano in Hawaii. These two are now the largest single telescopes in the world.
- The New Mexico State University runs the Apache Point Observatory in the Sacramento Mountains in New Mexico. This observatory is involved in the APOLLO Project. APOLLO stands for Apache Point Observatory Lunar Laser-ranging Operation. Starting in 1969, American astronauts placed reflecting devices on the moon. The Russians used remote-control systems instead of astronauts to place more reflecting devices on the moon. By observing these devices, astronomers can make very accurate measurements of the distance between Earth and the moon. They have been making these measurements for more than 35 years.

The astronomers at Apache Point are also involved in a project called the Sloan Digital Sky Survey (SDSS). Their goal is to make a map of a quarter of the sky and discover information on about 100 million space objects.

Observatories in Space

Until the late 1950s, all of the observatories that astronomers used were located on Earth. But in 1957, the Soviet Union sent the satellite *Sputnik I* into space. It was about the size of a basketball and weighed only 83 kg (183 lb). *Sputnik* was the first human-made object to orbit Earth. Its launch was the beginning of the Space Age.

The United States sent its first satellite into space about four months after *Sputnik* was launched. The satellite was called *Explorer I*. It was the first space observatory, even though it didn't have a telescope. *Explorer I* contained scientific equipment that helped scientists discover the rings of radioactive particles that surround our planet.

The higher a telescope is, the less atmosphere observers must look through. A telescope floating in space, where there is no atmosphere at all, is ideal! After scientists succeeded at launching objects into orbit, astronomers wasted no time sending up telescopes. They have now been sending telescopes into space for more than 40 years.

Some Interesting Space Observatories

- The British launched the first orbiting telescopes beginning in the early 1960s. Their *Ariel* series studied the sun.
- In 1968, the United States space agency, NASA, launched the first successful OAO (Orbiting Astronomical Observatory). It contained telescopes and other instruments.
- The IUE (International Ultraviolet Explorer) was launched by NASA and two European space agencies in 1978. It was expected to work for 3 years, but it lasted 18 years before it was finally turned off.
- IRAS (InfraRed Astronomical Satellite) used optical instruments to study the entire sky. Through the information provided by these instruments, scientists discovered rings of dust and gas around nearby stars. IRAS was launched in January 1983 and stopped working in October of that year.
- The Hubble Space Telescope was launched by NASA in 1990. It is a reflecting telescope named after Edwin Hubble, a famous astronomer who made important discoveries about galaxies in the first half of the 1900s. It is still orbiting Earth, taking photographs of distant objects in space. After the Hubble telescope was launched, astronomers realized that its mirror had not been made properly. Because of this, its photographs were distorted. In 1993, the space shuttle *Endeavor* was launched into space to fix the problem. The astronauts did their job well, and the problem was corrected. There are plans to replace the Hubble telescope with the James Webb Space Telescope sometime after the year 2009.
- The Canadian satellite SCISAT-1, which stands for SCIence SATellite Number 1, was launched in August 2003. SCISAT-1 uses many different instruments, including optical instruments, to gather data about the upper atmosphere of Earth.

The Hubble Space Telescope

- On June 30, 2003, Canada launched the smallest space telescope yet, called MOST. MOST stands for Microvariability and Oscillations of STars. MOST is the size of a large suitcase and is expected to be useful for about five years. One of the missions of MOST is to look for light reflecting off planets that are orbiting distant stars.

SOME THINGS ASTRONOMERS HAVE LEARNED . . .

. . . About Our Solar System

The solar system is the term we use to describe everything that revolves around the sun. Thanks to astronomers, we know that our solar system includes planets, moons, asteroids, and comets. We also know that there are other solar systems in the universe that contain planets.

. . . About the Sun

Astronomers discovered that the sun is a huge ball of burning gas in the center of our solar system. Thanks to them, we know that the sun is actually an average-sized star. There are many stars that are bigger and brighter than the sun. Our sun seems brighter than other stars only because it is so much closer to us.

The sun's heat and light enable life on Earth to exist. Astronomers know that the sun is about 4.5 billion years old and will probably keep shining for another 7 billion years or so.

Early cultures thought that the sun, moon, and stars all revolved around Earth. Because of astronomers, we know that the sun is really the center of our solar system. The sun's gravity keeps everything from flying off into space.

Astronomers use special filters to examine the sun. These filters break sunlight into different colors. Different kinds of gases produce different combinations of colors. By examining the colors, astronomers have figured out that the sun consists almost entirely of two gases, hydrogen and helium.

. . . About the Moon

Astronomers have taught us that the moon does not produce light. We can see only the part of it that is reflecting the light of the sun. Sometimes the sun lights up the moon's face completely. We call that a full moon. Other times, only part of the moon or none of it can be seen. Astronomers know that the moon's phases are caused by the relative positions of the moon, the sun, and Earth.

People have always been captivated by the moon. There have been myths, stories, and superstitions about it throughout recorded history. The Sumerians had a different god for each of three different ways that the moon looked. The ancient Chinese believed that there were 12 different moons, all with the same mother. This mother would take one moon each month on a chariot ride across the sky.

Today we know that the moon is made mostly of rock and metal. We know that it has no atmosphere. We also know that it has no liquid water and that nothing lives on the moon.

Did you know that other planets have moons, too? Mars has two of them. So far, astronomers have found more than 60 moons for Jupiter. Some scientists think that there might be primitive forms of life on Europa, one of Jupiter's moons.

. . . About Planets

A planet is a large body that orbits our sun or another star. Planets do not make light on their own. Whatever brightness they have is from reflected light, just as the moon's light is. The planet we know best, of course, is Earth. Other planets in our solar system are Mercury, Venus, Mars, Jupiter, Saturn, Uranus, Neptune, and Pluto. These names are taken from the names of ancient Greek and Roman gods because it was once believed that the gods lived in the sky. Some planets, for example, Mars and Venus, can at times be seen by the naked eye. They look like bright stars. To see them well, you need a good telescope.

Each planet is different from the others. Mercury is very hot on one side and very cold on the other. Saturn has rings of dust and rock floating around it. Some people think there might be very primitive life on Mars, or at least that there once was. The only planet on which we know for sure that life exists is Earth.

. . . About Asteroids

Asteroids are pieces of rock that orbit the sun. They are much smaller than planets. Sometimes they are called planetoids. The largest asteroid we know about is a little more than 600 miles long.

Asteroids have hit Earth in the past. Some scientists think that dinosaurs were wiped out when an asteroid hit Mexico about 65 million years ago. In 1908, a huge explosion killed thousands of reindeer in a remote part of Siberia. Astronomers think that the explosion was caused by an asteroid exploding just before it hit the ground.

Think how terrible it would have been if that asteroid had landed near people. Astronomers and other scientists have thought about this. Since 1995, the United States government has used an observatory in Hawaiʻi to keep track of asteroids and decide whether any of them could be a danger to people. They can see where an asteroid is headed for years before it gets there. If scientists discovered that a big asteroid was headed our way, they would have enough time to destroy it.

Once in a while, a new asteroid is discovered. Sometimes an amateur astronomer discovers one. The International Astronomical Union confirms the discovery of each asteroid and lets whoever discovers it choose the asteroid's name. If you discovered an asteroid, what would you name it?

. . . About Comets

The word *comet* comes from the Latin words *stella cometa*, meaning "hairy star." A comet is a big chunk of rock and ice that orbits the sun. Most of the time, a comet is far from the sun. When it gets close to the sun, some of its ice melts and evaporates, turning into gas. This gas forms a long glowing tail that follows the comet. Comets can sometimes be seen from Earth without a telescope. Astronomers know of about 2000 comets in our solar system. Many were first spotted in ancient times, but several hundred of them were not found until after telescopes were invented.

The most famous comet of all is Halley's Comet. It was first noticed by Chinese astronomers about 2200 years ago. In the 1700s, an English astronomer named Edmond Halley predicted when the comet would appear. When the comet appeared at the time that Halley had predicted, other astronomers named the comet after him. Halley's Comet can be seen by the naked eye from Earth for a few days every 76 years or so. It last came through our solar system in 1986 and is due back in 2061.

. . . About Galaxies

A galaxy is an enormous group of stars. About 5000 different stars are bright enough to be seen without a telescope. These stars all belong to our own galaxy, the Milky Way. Hundreds of thousands of other stars can be seen with a telescope, even a smaller one. With a large telescope in an observatory, astronomers can see millions of galaxies. And each galaxy contains billions of stars.

. . . About the Universe

Astronomers use the word *universe* when they mean "everything there is." Most of them think that about 13 or 14 billion years ago, everything in the universe was packed into one very small, hot mass. An event called the "Big Bang" is thought to have caused this mass to explode. In this explosion, the universe burst out, and it has been expanding ever since.

What Is Next?

It is impossible to predict what discoveries astronomers might make in the future. Many things that we now know about space would not have been believed by people in the past. One thing is certain, though—as long as there are people, there will be someone to wonder about the universe and try to learn more about it. Even if they have to stand on tiptoe.

The Milky Way—our galaxy